How to be Brilliant at

RECORDING IN GEOGRAPHY

Sue Lloyd

Brilliant Publications

We hope you and your class enjoy using this book. Other books in the series include:

History title
How to be Brilliant at Recording in History 1 897675 22 4

Science titles
How to be Brilliant at Recording in Science 1 897675 10 0
How to be Brilliant at Science Investigations 1 897675 11 9
How to be Brilliant at Materials 1 897675 12 7
How to be Brilliant at Electricity, Light and Sound 1 897675 13 5
How to be Brilliant at Living Things 1 897675 66 6

English titles
How to be Brilliant at Writing Stories 1 897675 00 3
How to be Brilliant at Writing Poetry 1 897675 01 1
How to be Brilliant at Grammar 1 897675 02 X
How to be Brilliant at Making Books 1 897675 03 8
How to be Brilliant at Spelling 1 897675 08 9
How to be Brilliant at Reading 1 897675 09 7
How to be Brilliant at Word Puzzles 1 897675 88 7

Maths titles
How to be Brilliant at Using a Calculator 1 897675 04 6
How to be Brilliant at Algebra 1 897675 05 4
How to be Brilliant at Numbers 1 897675 06 2
How to be Brilliant at Shape and Space 1 897675 07 0
How to be Brilliant at Mental Arithmetic 1 897675 21 6

Other titles
How to be Brilliant at Christmas Time 1 897675 63 1

If you would like further information on these or other titles published by Brilliant Publications, please write to the address given below.

Published by Brilliant Publications,
1 Church View,
Sparrow Hall Farm,
Edlesborough,
Dunstable,
Bedfordshire
LU6 2ES

Sales and stock enquiries:
Tel: 01202 712910
Fax: 0845 1309300
e-mail:brilliant@bebc.co.uk
www.brilliantpublications.co.uk

General information enquiries:
Tel: 01525 229720

The name 'Brilliant Publications' and the logo are registered trademarks.

Written by Sue Lloyd
Illustrated by Kate Ford
Cover photograph by Martyn Chillmaid

Printed in the UK

© Sue Lloyd 1998
ISBN 1 897675 31 3

First published in 1998. Reprinted in 2005.
10 9 8 7 6 5 4 3 2

The right of Sue Lloyd to be identified as author of this work has been asserted by her in accordance with the Copyright, Designs and Patents Act 1988.

Contents

Introduction

How to be Brilliant at Recording in Geography contains 42 photocopiable worksheets designed to lead children through a range of geographical skills. The sheets are divided into six sections which reflect the Geographical Skills and Thematic Studies of the National Curriculum:

- skills
- map work
- rivers
- localities and settlements
- weather
- environmental change

The sheets have been arranged according to their main focus, but all the sheets cover work from other areas of the National Curriculum. For example, page 46 (Natural resources) is in the Environmental Change section, but can also be used for work on Localities and Settlements. The sheets have been designed to help children:

- investigate places and themes
- gain knowledge and understanding about places and themes
- ask geographical questions
- develop the ability to recognize and explain geographical patterns
- become aware of how places fit into a wider geographical context.

Most of the worksheets are self-explanatory, but we recommend that you read the teachers' notes before using the sheets. The notes will give you pointers and ideas for each sheet. These are only suggestions and there are many ways of adapting the material to suit the needs of individual children. The sheets may be used for individual, paired and group work or for whole class activities.

Using the worksheets

SKILLS

Questions and answers   **page 7**
- Some children will need more than one sheet.

Planning an investigation   **page 8**
- Suggested topics: the effects of a river on its landscape, how quarrying or mining changes a region, the opening of a local supermarket.

Mind map   **page 9**
- Can be used to assess children's knowledge before starting a new topic, or to assess their understanding at the end of a topic (or both).
- Can also be used as a structured brainstorm-type activity.
- Can be used to supplement page 48 (Input-output).

Picture this   **page 10**
- Children can use pictures from comics, magazines, brochures or from their own work.

Flow diagram   **page 11**
- Children can write any changes that have taken place inside the thick arrows.

Graph   **page 12**
Table   **page 13**
Charts   **page 14**
- Children should give their graph, table or chart a heading and label the axes on the graph or chart appropriately.

Field trip   **page 15**
- This sheet helps children to focus on the reasons for going on a field trip.
- Use the sheet to help recall experiences and information as part of follow-up in school.

My trail   **page 16**
- Suggested trails: around school, on a field trip, at home.

Special report   **page 17**
- Suggested reports: to stop a new road being built, to persuade local business people to develop a derelict area, to stop (or start) mining.

Campaign   **page 18**
- Can be used as a follow-on to page 17 (Special report).

Time to disagree   **page 19**
- Children can take opposing views, or explore the views of two other people.
- Can be used to supplement pages 17 and 18 (Special report and Campaign).

Review sheet   **page 20**
- Use the sheet to assess children's understanding.

MAP WORK

Map quiz   **page 21**
- Children can draw their own map, construct a key and set their own questions.

Where in the UK do you live?   **page 22**
Where in Europe do you live?   **page 23**
Where in the world do you live?   **page 24**
- Children could add links between different places.

RIVERS

Rivers, 1   **page 25**
- Children can work in groups to brainstorm 'river words'.

Rivers, 2   **page 26**
- Information from this sheet can be transferred to a computer database.

Water everywhere   **page 27**
- Children recall how we use and save water in our own environment.
- Can also be used to study water in other environments.

LOCALITIES

What is your address?   **page 28**
- Can be used for children's own details and/or contrasting localities.
- Children can draw themselves and the other person inside the smallest circles.

Fact file   **page 29**
- Information can be transferred to a computer database.

Comparing localities — page 30
- Children can draw or describe the localities in the large boxes.

Using my senses — page 31
- Suggested locations: in school, at the shops, at home, in the park.

If I were in ... — page 32
- Suggested locations: in a town, on a farm, on a beach, in a forest.

Where am I? — page 33
- Children can use their knowledge and understanding to formulate five clues and two diagrams/pictures. They then give the sheet to a friend to complete.

A day in the life — page 34
- The line is divided into 24 sections.
- Children could interview a person.
- The various activities could be colour-coded, eg: work, sleep, travel, leisure.

Building a home — page 35
- Children draw a home and write building materials used in the blocks around the edge.
- They draw lines from the materials to the building to show where the materials are used.
- This sheet can be used twice to compare different buildings.

On the move — page 36
- Can be used as an introduction to transport.
- Can also be used as part of the work on a settlement or locality.

Planning a holiday brochure — page 37
- Encourage children to use this sheet as a plan and not to include too much detail.

WEATHER
Weather diary — page 38
- There are several boxes to challenge children to think of as many weather variations as possible.
- Columns are blank to allow the teacher or child to choose which days to include.
- Children can use more than one symbol in each box.

Worries about the weather — page 39
- Focus on one locality or use as a general study of weather conditions.

Worldwide weather — page 40
- Can be used to compare localities
- Suggested chart headings: rainfall, temperature, natural disasters.

Site conditions — page 41
- Children can draw or describe the site, eg: under a tree, behind a shed, in the middle of the playground.
- Suggested headings for small columns: rainfall, temperature, sunshine, windspeed.

ENVIRONMENTAL CHANGE
The environment — page 42
- Suggested topics: local shopping precinct, town centre, river, farm.

Changes — page 43
- Focus on a specific area, eg: the Nile, the Amazon rainforest, a Caribbean island.
- Alternatively, focus on general topics, eg: rivers, cliffs, mountains, deserts.

Damage to our environment — page 44
- Can be used to study one area, eg: problems around school.
- Can also be used to study problems in larger areas, eg: desert or polar regions.

Improvements to our environment — page 45
- Long term benefits could be over 10, 20, 30 years or more.

Natural resources — page 46
- Suggestions for 'this is...' boxes: my desk, my trainer, my watch.
- Suggestions for ovals: it is made from..., it comes from..., I use it ..., did you know...?

Tour of inspection — page 47
- Can be used to look at damage around school, at home, or as part of a field trip.

Input – output — page 48
- Can be used to record our changing environments, eg:
  heavy rainfall ➔ flooding ➔ homes destroyed;
  factory farming ➔ battery hens ➔ more eggs;
  reservoir ➔ new lake ➔ more water for people;
  reservoir ➔ new lake ➔ people's homes flooded.
- Can be used to supplement page 9 (Mind map).

Questions and answers

I want to find out …

I will ask these questions:

This is how I will find the answers:

These are the answers:

How to be Brilliant at Recording in Geography
www.brilliantpublications.co.uk

Planning an investigation

I am investigating …

I will need …

I want to find the answers to these questions:

These are the answers I have found:

Mind map

I am thinking ...

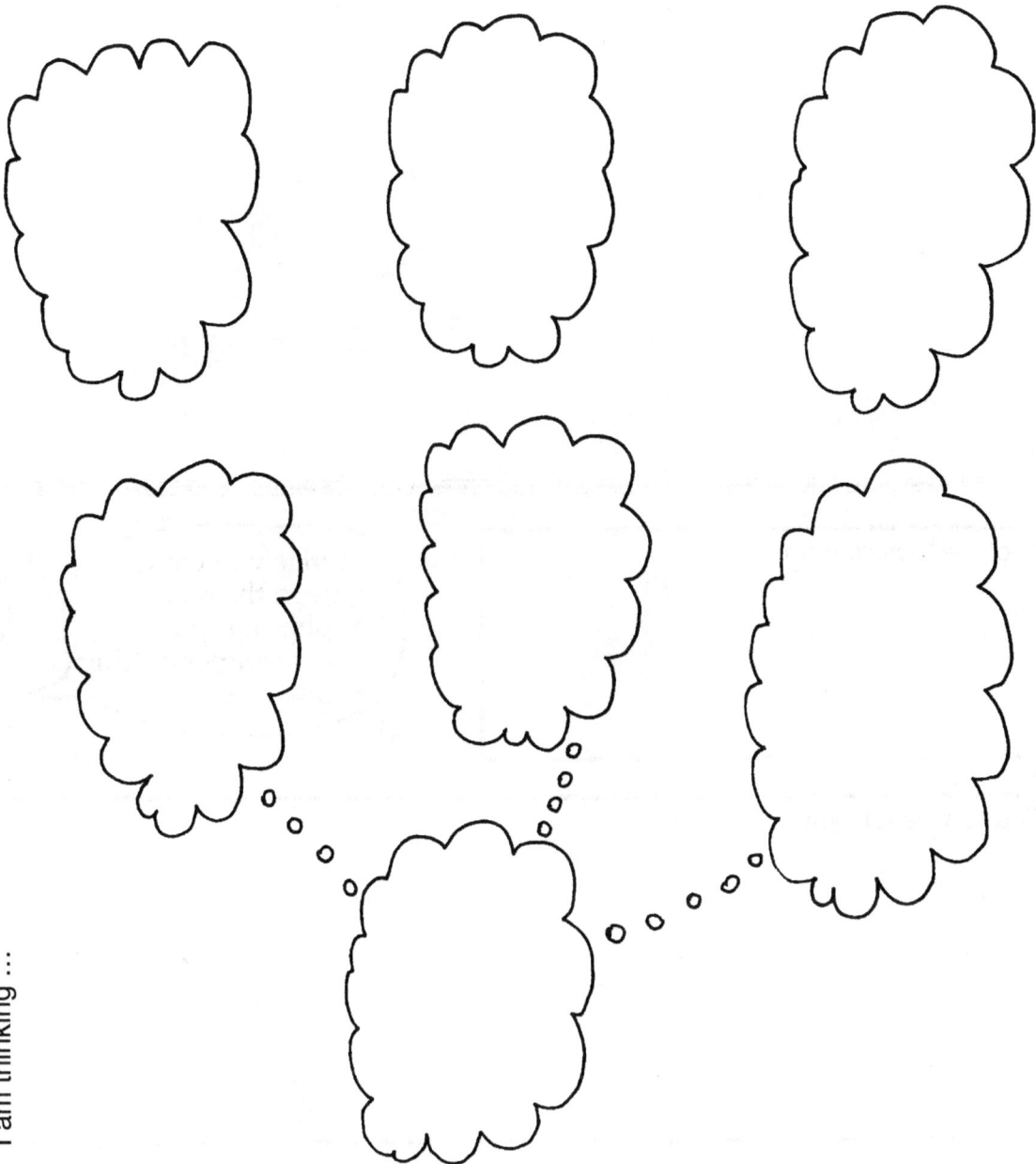

Try to link your ideas. When you get more ideas or information, add extra bubbles.

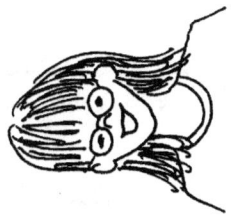

How to be Brilliant at Recording in Geography
www.brilliantpublications.co.uk

Picture this

This is a picture of ...

I got the picture from ...

Things you could
glue in the box:
• photograph
• newspaper cutting
• postcard.

The picture tells me ...

Flow diagram

This flow diagram is about …

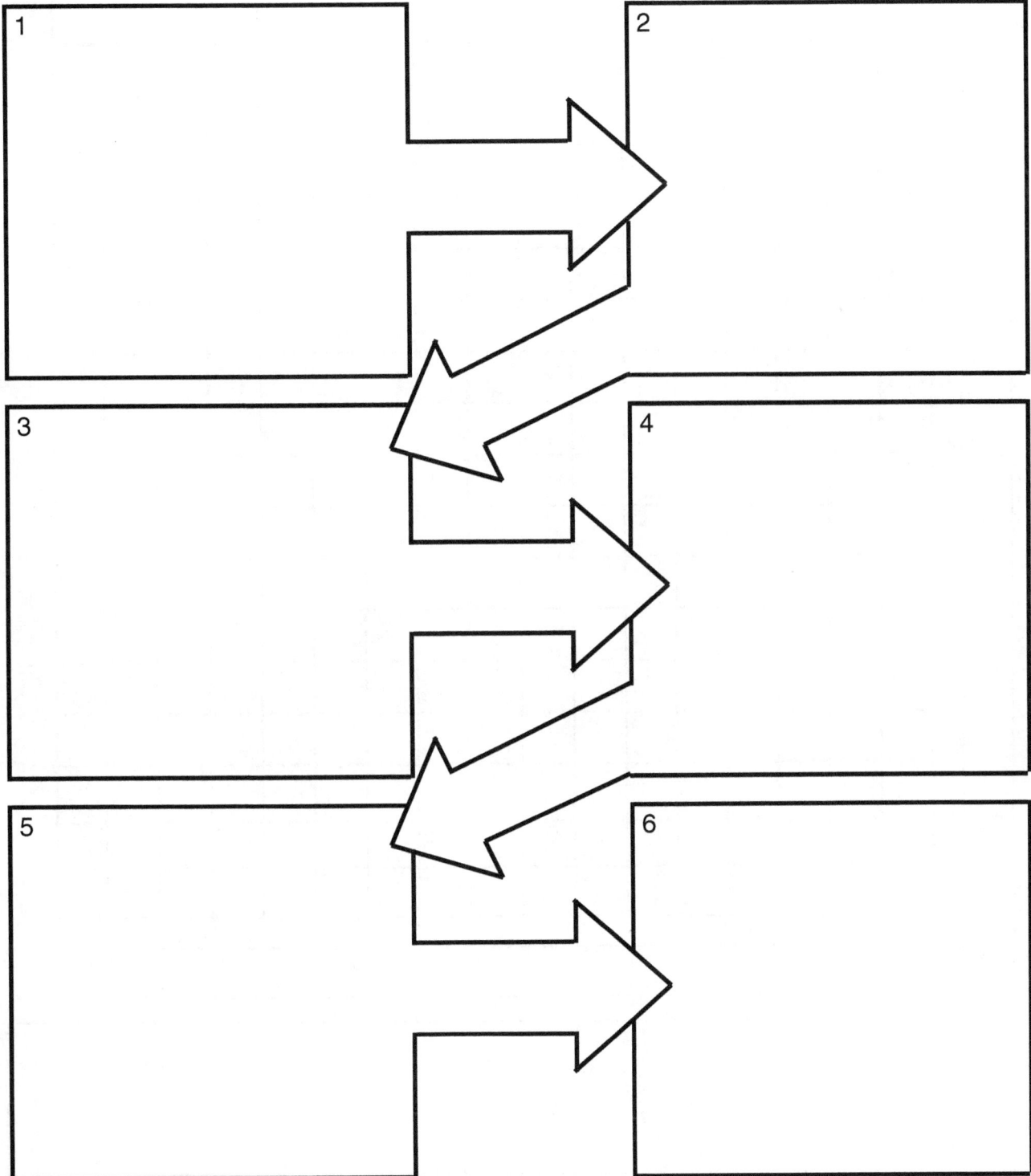

1

2

3

4

5

6

How to be Brilliant at Recording in Geography
www.brilliantpublications.co.uk

Graph

This is the title of my graph:

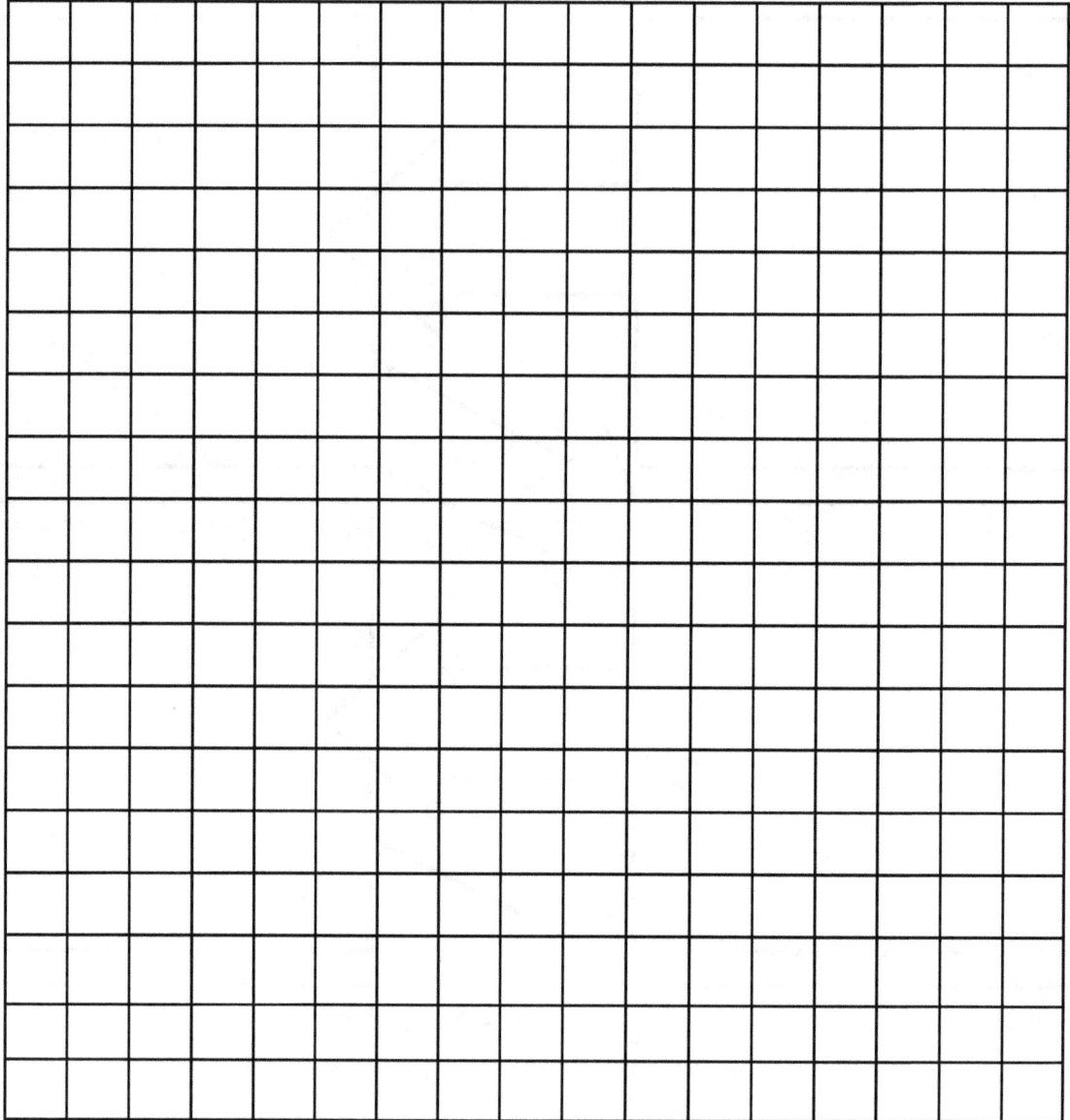

My graph shows …

How to be Brilliant at Recording in Geography
www.brilliantpublications.co.uk

Table

This is the title of my table:						

My table shows …

Charts

This chart is about ...

This chart is about ...

The charts tell me ...

Field trip

Before the trip

I am going on a field trip to…

The aims of the trip are …

I will need to take …

After the trip

I found out…

The best part of the trip was …

My trail

Start at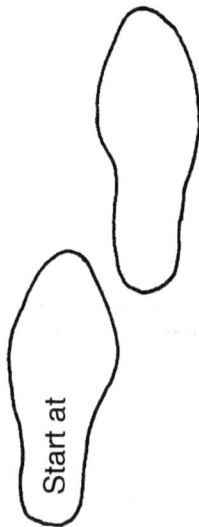

1 Directions:

Interesting points:

2 Directions:

Interesting points:

3 Directions:

Interesting points:

Use this space for a map or picture.

4 Directions:

Interesting points:

5 Directions:

Interesting points:

Finish at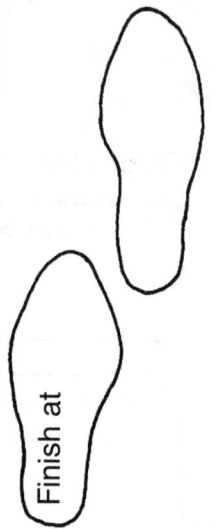

Special report

This is what _____ thinks:

This is what _____ thinks.

After reading these opinions, I think …

Reporter's name: _____

This report is …

In my opinion …

Campaign

This is what my publicity will look like:

My campaign is ...

These are my reasons ...

I will ...

Time to disagree

This is a disagreement about …

Arguments for	Arguments against

This is what I think:

This is what_____ thinks:

This is what we agreed:

Review sheet

I have finished learning about …

This is what I can remember …

Use these boxes for drawings or maps.

The most enjoyable work was …

The least enjoyable work was …

Map quiz

? ? ? ? ? ? ? ? ? ? ? ? ? ? ? ?

Questions

1
2
3
4
5
6

Answers

1
2
3
4
5
6

Key

Where in the UK do you live?

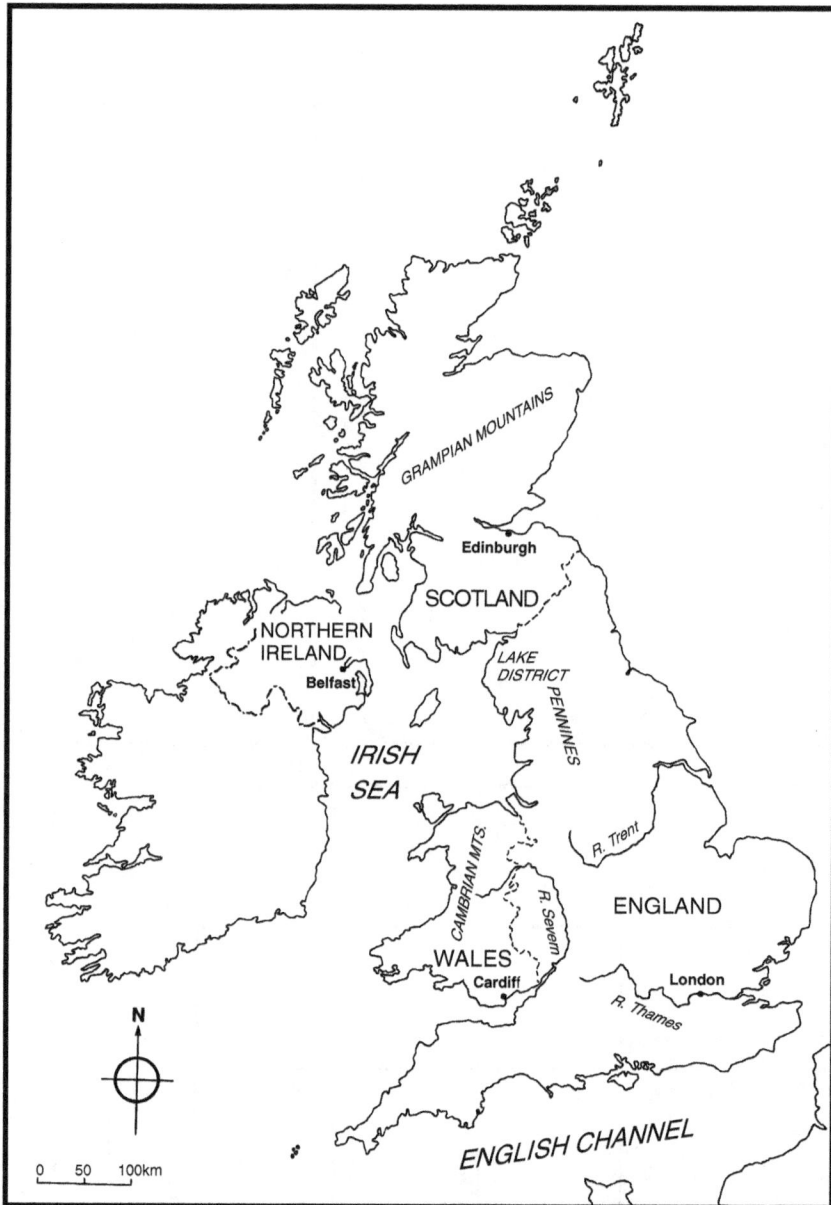

Place	Distance from my home	Direction

Use the scale bar and compass points to help you complete the table.

Where in Europe do you live?

Map of Europe showing: Iceland, Republic of Ireland, Dublin, United Kingdom, London, North Sea, Germany, Berlin, Paris, France, Alps, Italy, Rome, Madrid, Spain, Mediterranean Sea. Compass marked N. Scale: 0 — 500km.

Place	Distance from my home	Direction

How to be Brilliant at Recording in Geography
www.brilliantpublications.co.uk

Where in the world do you live?

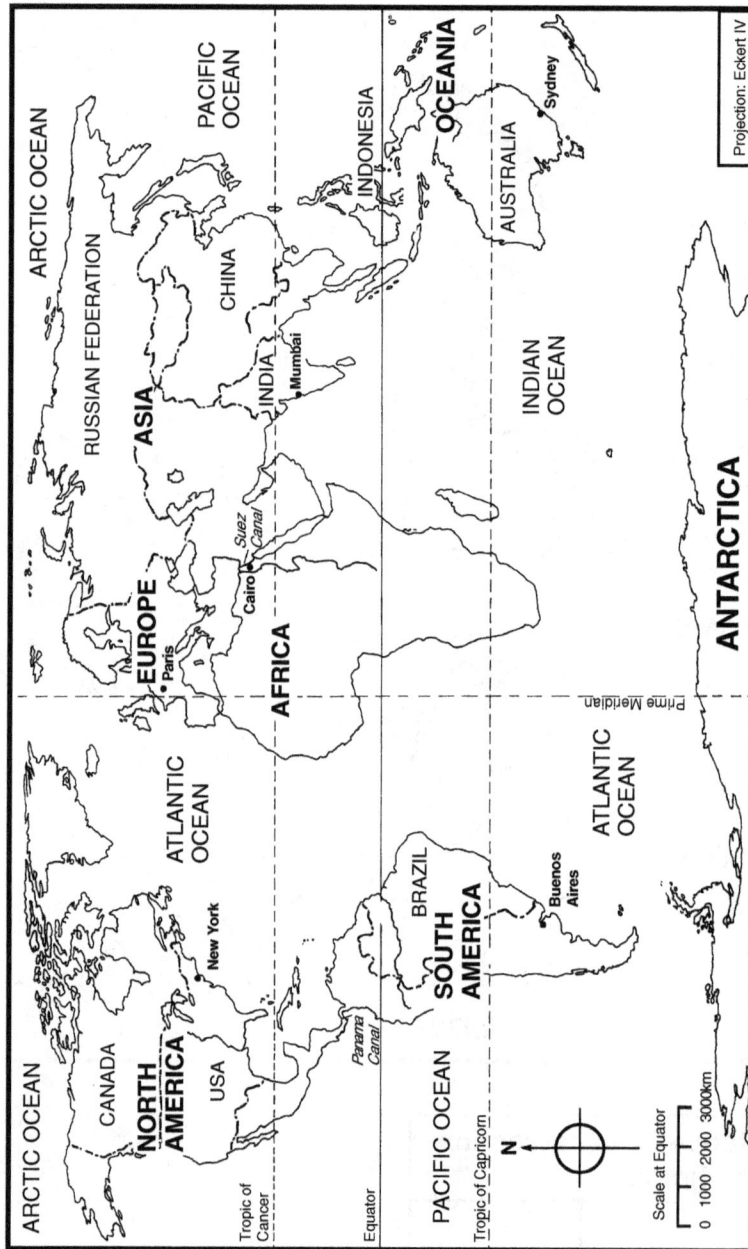

Place	Distance from my home	Direction

ARCTIC OCEAN

ARCTIC OCEAN

RUSSIAN FEDERATION

PACIFIC OCEAN

ASIA

CHINA

INDONESIA

OCEANIA

Sydney

AUSTRALIA

INDIAN OCEAN

Mumbai

INDIA

Suez Canal

Cairo

EUROPE

Paris

AFRICA

ANTARCTICA

Prime Meridian

ATLANTIC OCEAN

CANADA

NORTH AMERICA

USA

New York

Panama Canal

BRAZIL

SOUTH AMERICA

Buenos Aires

ATLANTIC OCEAN

Tropic of Cancer

Equator

PACIFIC OCEAN

Tropic of Capricorn

N

Scale at Equator

0 1000 2000 3000km

Projection: Eckert IV

Rivers, 1

river words

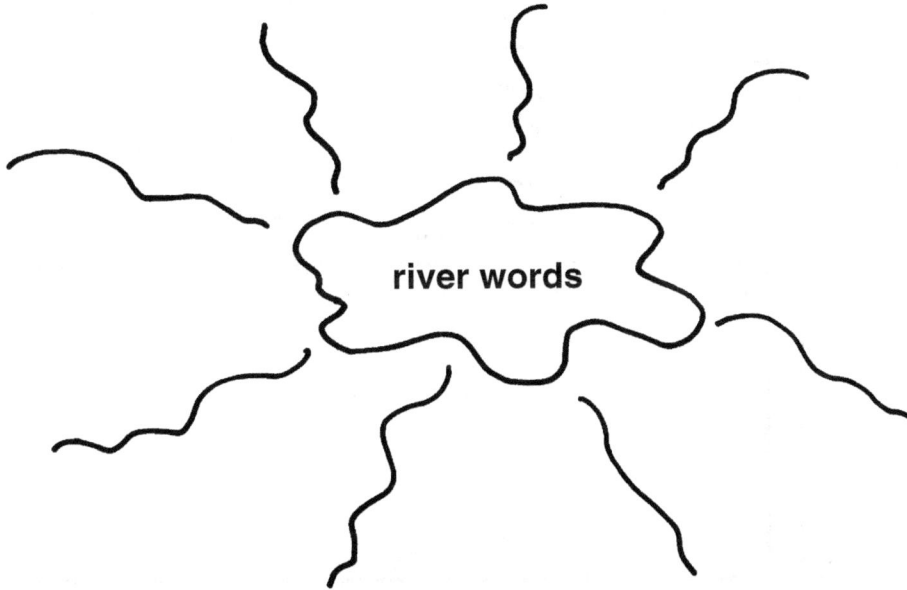

Use an ordnance survey map.
Draw water features here.

Explain their meaning here.

water feature	meaning

Rivers, 2

I am finding out about the river …

_____

Length:

It flows through …

These settlements are on the river:

People use it in these ways:

Did you know?

How to be Brilliant at Recording in Geography
www.brilliantpublications.co.uk

Water everywhere

This is how we use water.

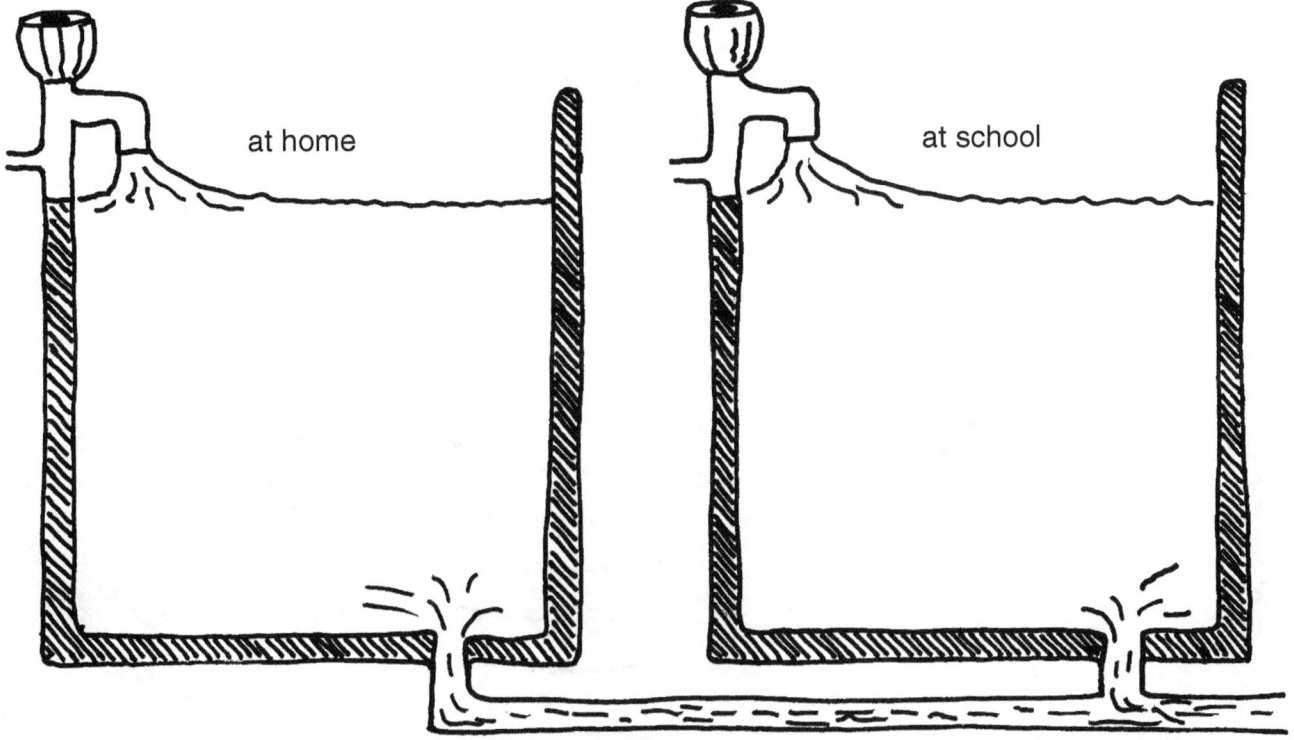

at home

at school

This is how we save water.

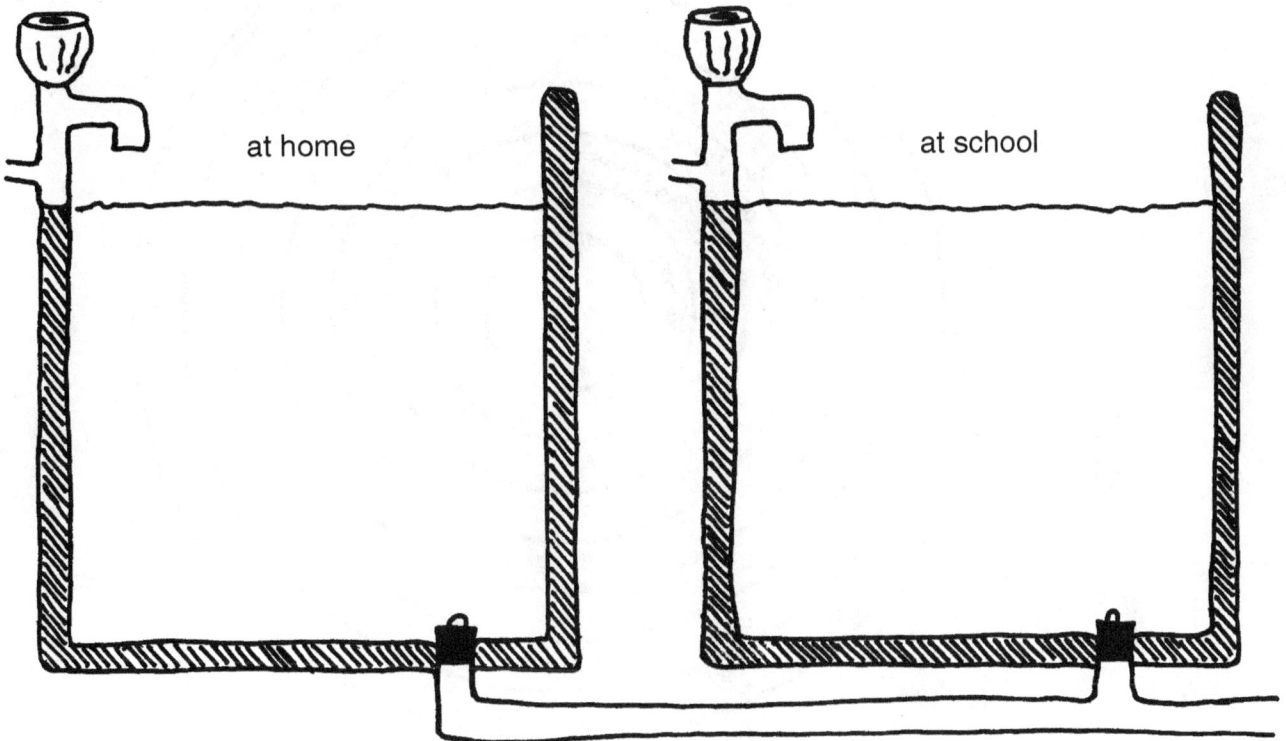

at home

at school

What is your address?

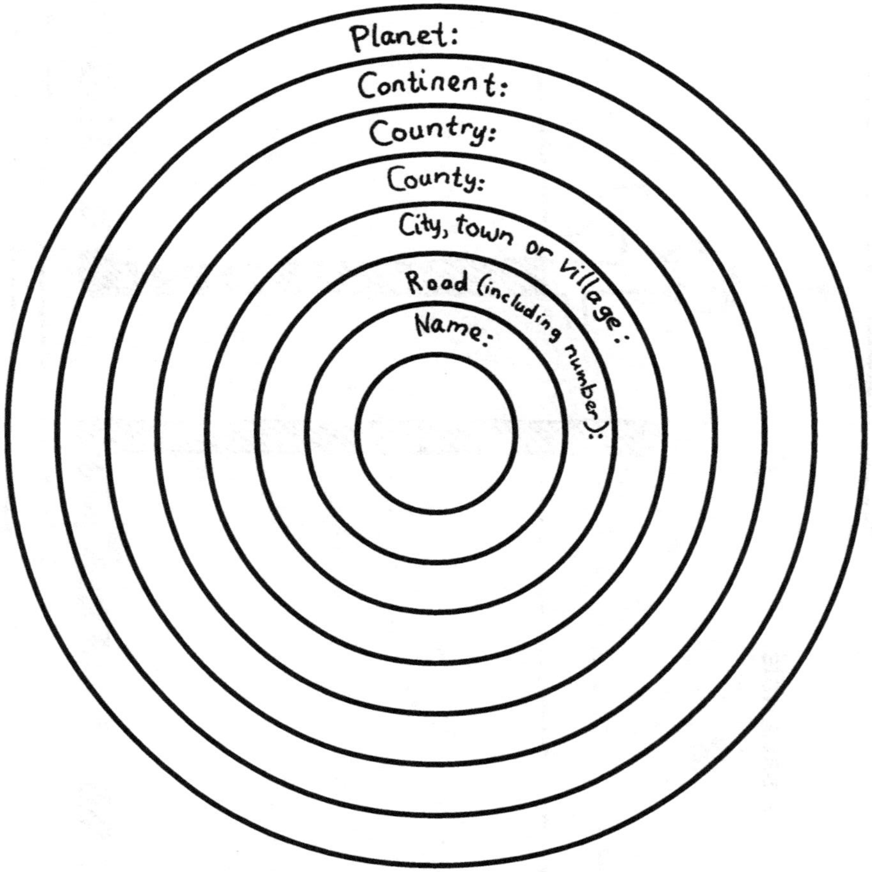
Planet:
Continent:
Country:
County:
City, town or village:
Road (including number):
Name:

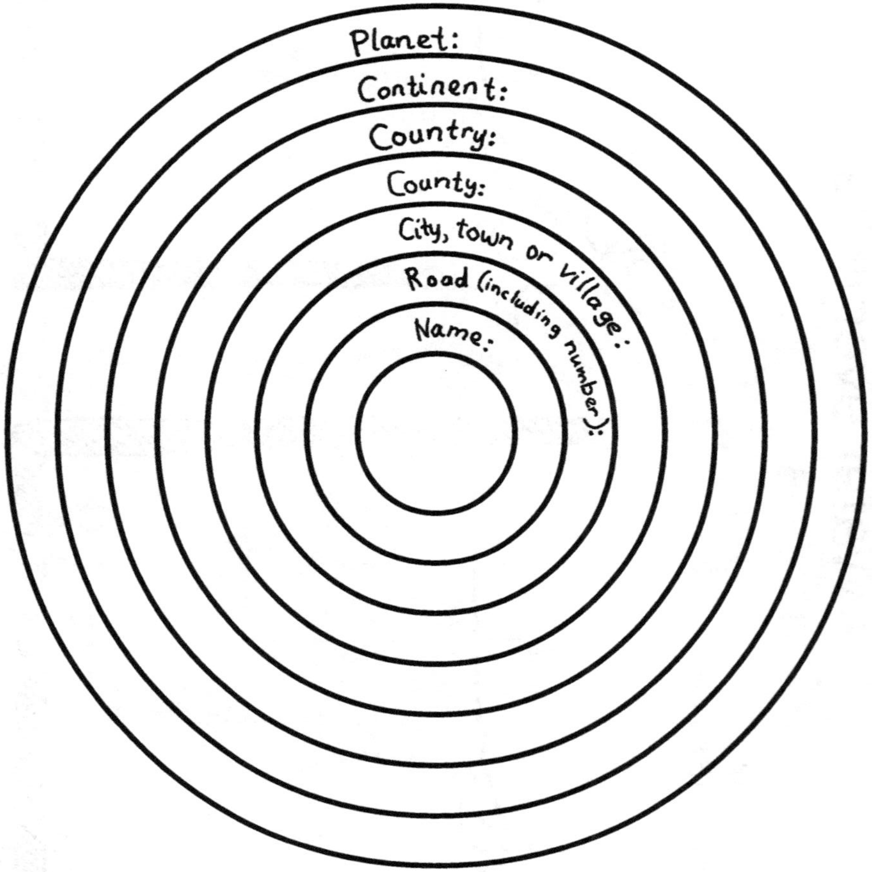
Planet:
Continent:
Country:
County:
City, town or village:
Road (including number):
Name:

Fact file

I am finding out about ... _____

Climate:

Population:

Language:

Landscape:

Agriculture:

Manufacturing:

Did you know?

Comparing localities

I am comparing …

_____ with _____

This is what it is like	This is what it is like

The differences are …

The similarities are …

The locality I prefer is …

I prefer it because …

Using my senses

I can use my senses to compare places.

Sense \ Place		
See		
Hear		
Smell		
Touch		
Taste		

The differences are …

The similarities are …

How to be Brilliant at Recording in Geography
www.brilliantpublications.co.uk

If I were in …

I might …

see

hear

touch
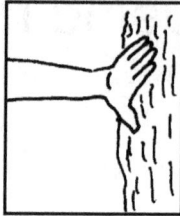

smell

taste

If I were in …

Draw or describe your location.

These are the things I might do:

Where am I?

Clues

1
2
3
4
5

Use these boxes to draw picture clues, charts or any other clues you can think of. See if a friend can guess where you are.

Name _____

I think you are in …

I think this because …

A day in the life

This is a day in the life of …

Time

Total 24 hours

Building a home

Draw a picture of a house. Write what building materials are used in the blocks around the edge.

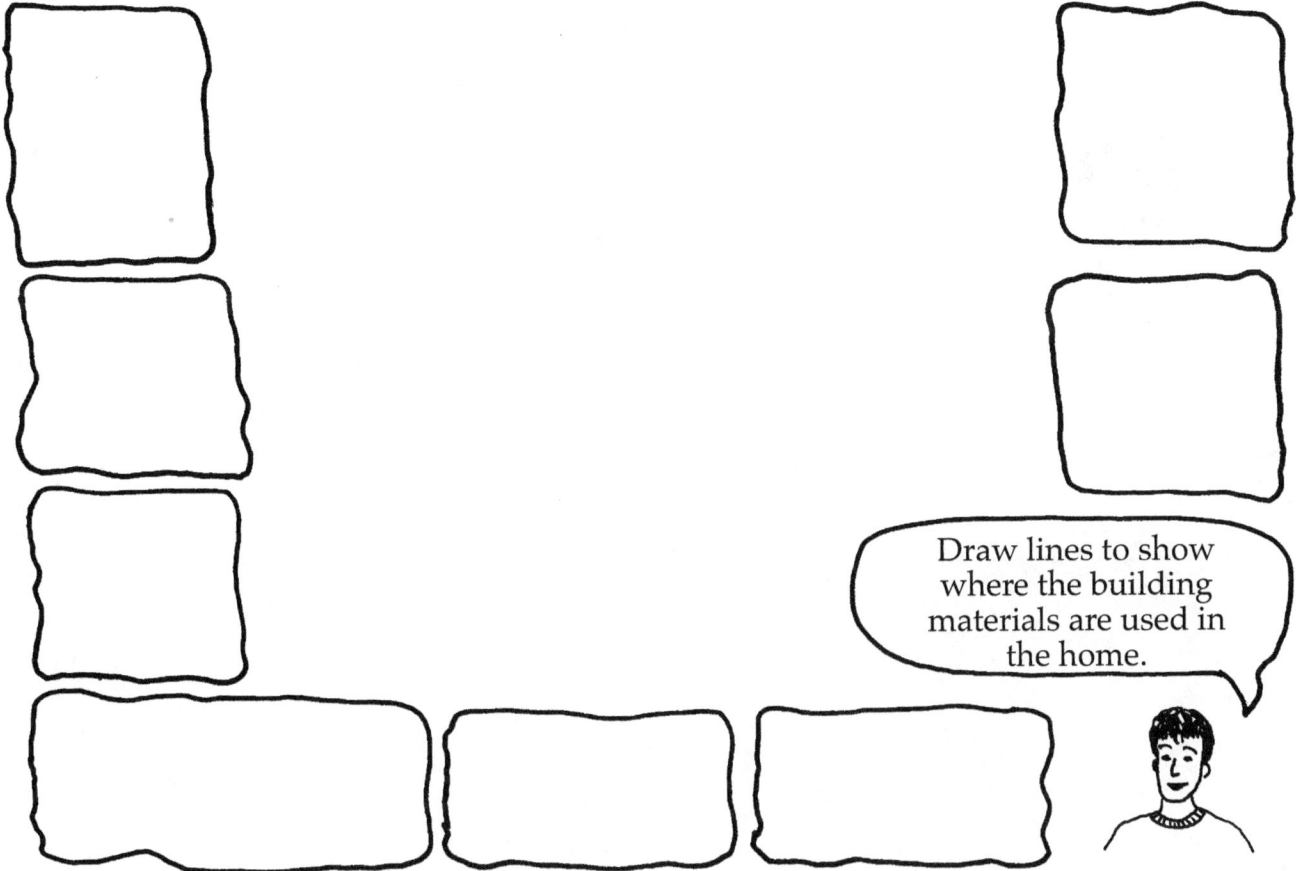

Draw lines to show where the building materials are used in the home.

Material	Use	This material was used because ...

How to be Brilliant at Recording in Geography
www.brilliantpublications.co.uk

On the move

What kinds
of transport
can you
think of?

In _____ people use …

Transport	Used for

Planning a holiday brochure

My brochure will be about …

I may include this information:

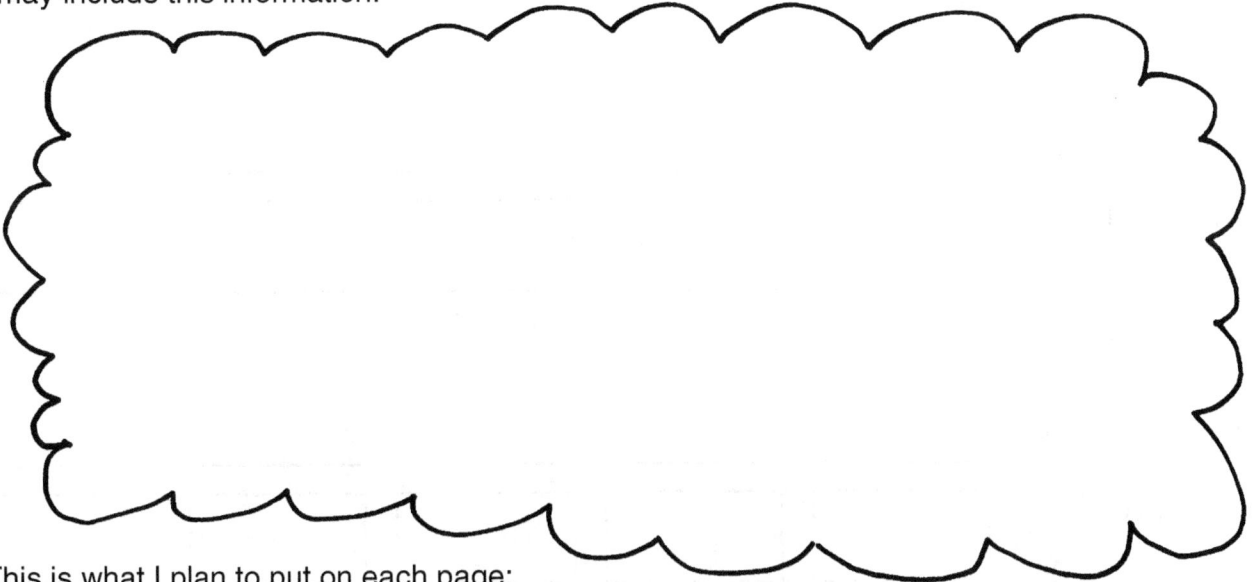

This is what I plan to put on each page:

1

2

3

4

5

6

Here is my design for
the front cover.

Don't go into much detail.
Use this sheet to plan your
work.

This is what _____
thinks of my ideas …

Weather diary

Weather condition _____ _____ _____ _____ _____

My symbol

Weather condition _____ _____ _____ _____ _____

My symbol

Weather condition _____ _____ _____ _____ _____

My symbol

I will use my symbols to complete this diary.

Day / Time					

Worries about the weather

I am looking at how the weather affects people's lives.

Weather condition	Problems	Possible solutions

Worldwide weather

I am finding out about weather conditions in these locations:

Location		

Location		

Location		

Site conditions

I am discovering how site conditions affect the weather.

Site 1

Description:

Site 2

Description:

Site 3

Description:

Site 4

Description:

This is what I have discovered:

How will you record your results?

I may use a graph or a bar chart.

The environment

I am finding out about …

It is being helped by …

It is being spoiled by …

I think …

This is what _____ thinks:

This is what _____ thinks:

After listening to other people's opinions, I think …

Changes

I am looking at changes …

Description:

Sketch or map:

These changes have taken place:

They have taken place because …

Advantages:

Disadvantages:

I think …

How to be Brilliant at Recording in Geography
www.brilliantpublications.co.uk

Damage to our environment

This is how we can damage our environment.

Location	Damage	My solutions

The greatest damage is …

I think this because …

My best solution is …

I think this because …

Improvements to our environment

This is how we can improve our environment.

Location	Improvement	My solutions

The most important improvement is …

I think this because …

These are the short term benefits:

These are the long term benefits:

The most important benefit is …

I think this because …

Natural resources

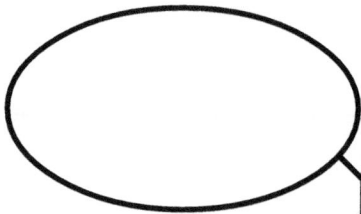

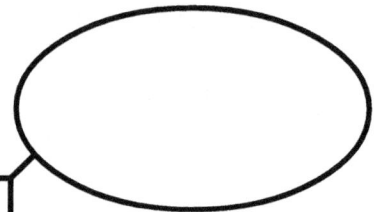

This is …

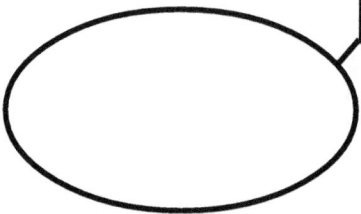

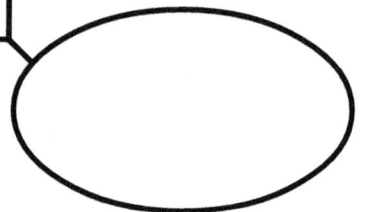

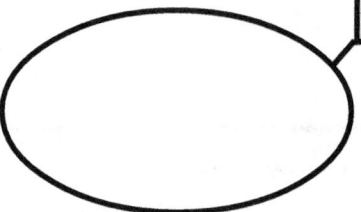

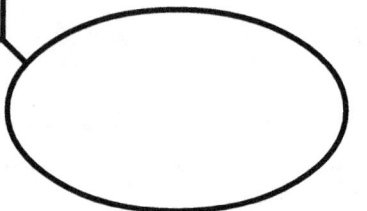

I got my information …

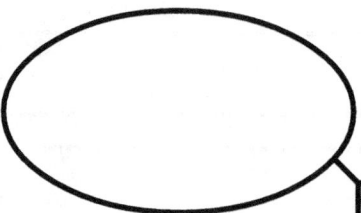

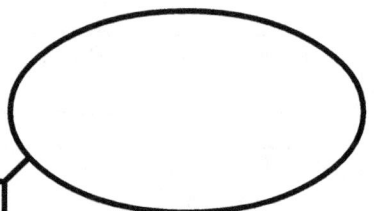

This is …

I got my information …

Tour of inspection

On this tour I will be inspecting …

I will start at …

Location	Damage	Caused by		Repairs
		People	**Nature**	

The worst damage is …

I think this because …

My ideas for preventing damage are …

Input – output

Input	What happens	Output

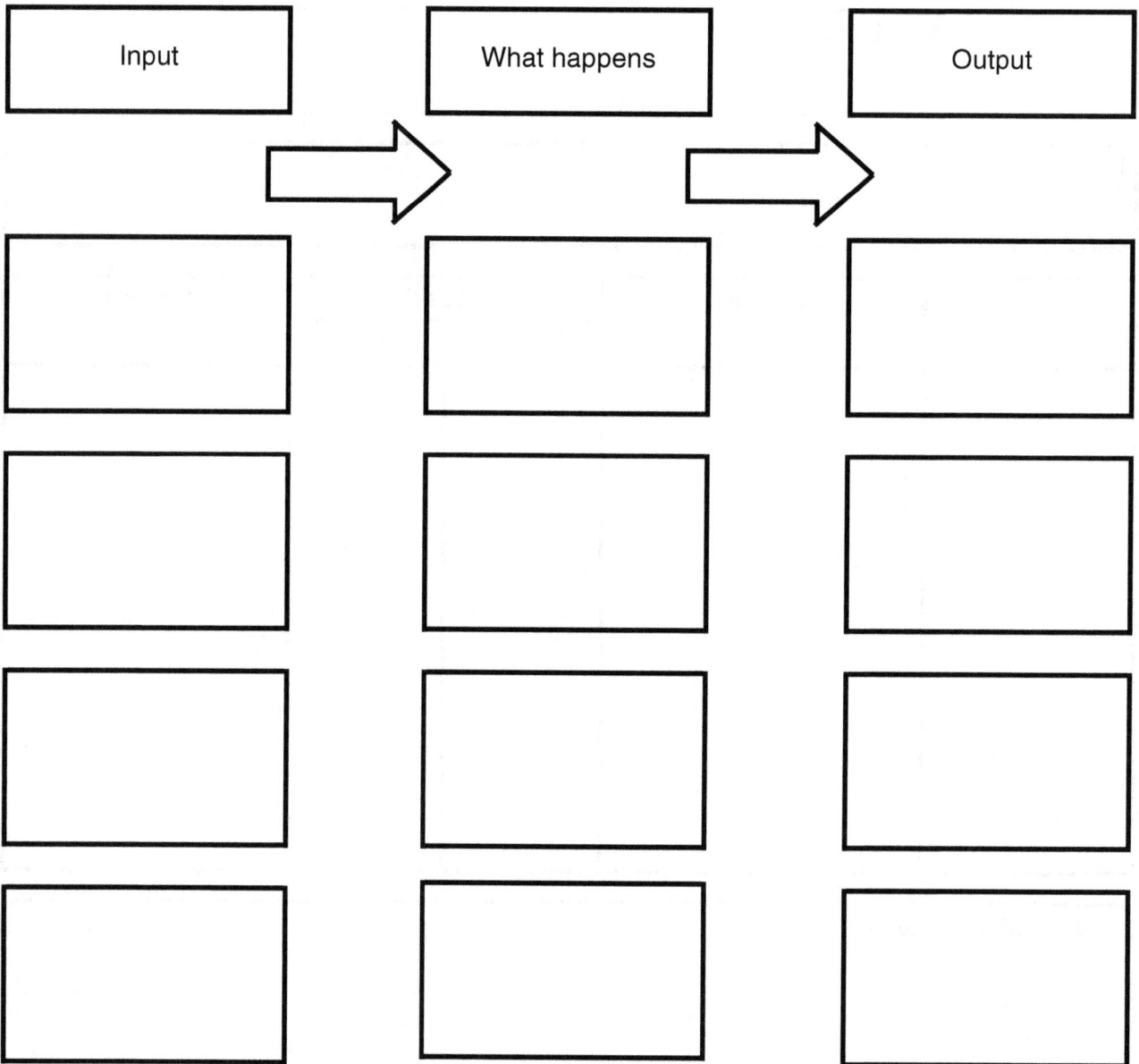

These inputs help:

These inputs do not help:

Also from Brilliant Publications

How to be Brilliant at Recording in History

SUE LLOYD

These generic writing frames help to develop historical skills and can be used to support and supplement any history topic. The book contains over 40 photocopiable worksheets focusing on chronology, developing historical knowledge and understanding, ways of interpreting history, historical enquiry and ways that pupils might organize and communicate what they have learned. Teachers' notes give pointers and ideas for each sheet.

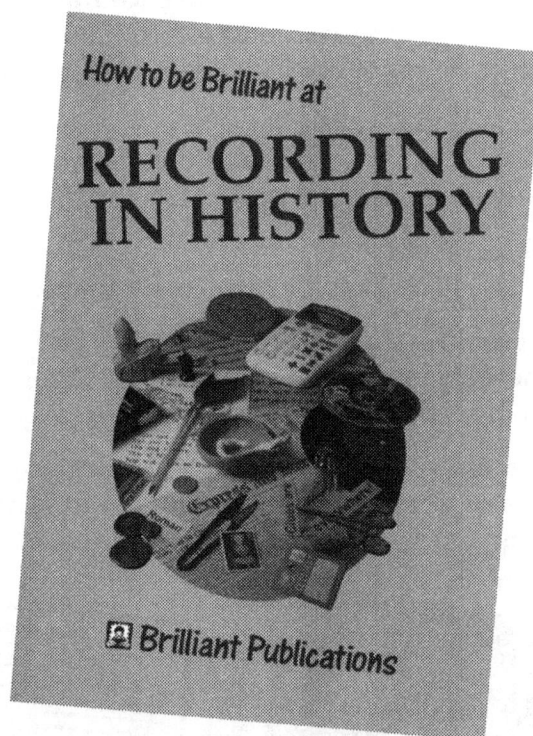

How to be Brilliant at Recording in Science

NEIL BURTON

40 photocopiable generic writing frames covering all aspects of investigations from planning through to analysing results.

The worksheets in this book are subdivided into six sections:

Teacher Sheets

The two sheets in this section are designed to assist the teacher in planning for pupil led investigations and to record assessments of how well individual children are meeting the learning objectives being set for them.

Thinking and planning

These sheets are designed to encourage children to think about the activity before they attempt it; to discuss and record ideas, predictions and hypotheses; and to concentrate on the planning of particular aspects of the science activity.

Recording results

These sheets provide structured formats that are sufficiently flexible to be used in a very wide range of situations, in conjunction with sheets from the previous section.

Observation

These sheets provide formats for the recording, ordering and sorting of observations.

Presenting findings

These sheets provide a wide range of formats suitable for presenting most forms of data that can be collected from investigations.

Content specific

This section contains sheets for the recording of ideas, observations and measurements about particular scientific content areas.

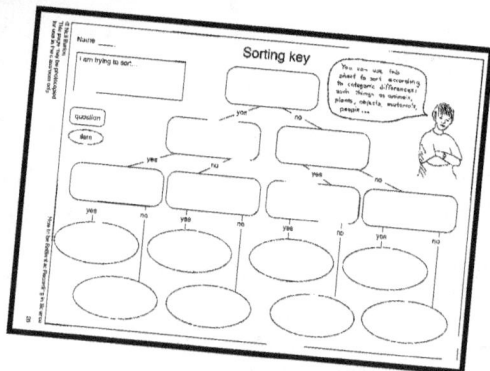

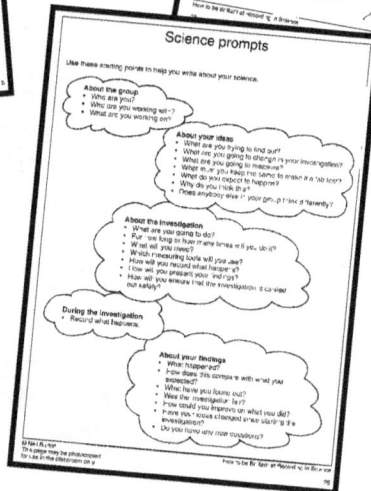

Other titles in the 'How to be Brilliant at...' series

Science Titles
How to be Brilliant at Electricity Light & Sound
How to be Brilliant at Living Things
How to be Brilliant at Materials
How to be Brilliant at Recording in Science
How to be Brilliant at Science Investigations

History Title
How to be Brilliant at Recording in History

Maths Titles
How to be Brilliant at Algebra
How to be Brilliant at Mental Arithmetic
How to be Brilliant at Numbers
How to be Brilliant at Shape & Space
How to be Brilliant at Using a Calculator

English Titles
How to be Brilliant at Grammar
How to be Brilliant at Making Books
How to be Brilliant at Reading
How to be Brilliant at Spelling
How to be Brilliant at Word Puzzles
How to be Brilliant at Writing Poetry
How to be Brilliant at Writing Stories

Other Title
How to be Brilliant at Christmas Time

You can find out more about our books on our secure website:

www.brilliantpublications.co.uk